Rainwater Harvesting for Your Homesteads

How to Collect, Store, and Use Rainwater Effectively

NORBERTO M. WRAY

Copyright © 2024 by Norberto M. Wray

All rights reserved. No part of this publication may be reproduced, distributed, or transmitted in any form or by any means, including photocopying, recording, or other electronic or mechanical methods, without the prior written permission of the publisher, except in the case of brief quotations embodied in critical reviews and certain other noncommercial uses permitted by copyright law.

Table of Contents

INTRODUCTION 6
    The Importance of Rainwater Harvesting 7
    Benefits for farmers 8
    Environmental impact 10
    Economic advantage 11

Chapter 1: Understanding Rainwater Harvesting 14
    1.1. What is Rainwater Harvesting? 16
    1.2. Definition and Overview 17
    1.3. Historical Context 19
    1.4. Benefits of Rainwater Harvesting for Farms 20

Chapter 2: Planning Your Rainwater Harvesting System 22
    2.1. Assessing Your Water Needs 24
    2.2. Estimating Water Usage 25
    2.3. Seasonal Variations 27
    2.4. Site Assessment 28

Chapter 3: Components of a Rainwater Harvesting System 31
    3.1. Catchment Areas 34
    3.2. Gutters and Downspouts 39
    3.3. Filtration Systems 43
    3.4. Storage Solutions 45
    3.5. Distribution Systems 47

Chapter 4: Designing and Building Your System 49
   4.1. System Design Principles 50
   4.2. Step-by-Step Installation Guide 55
   4.3. Common Challenges and Solutions 57

Chapter 5: Water Quality and Treatment 59
   5.1. Ensuring Water Quality 62
   5.2. Testing and Monitoring 63
   5.3. Common Contaminants and Solutions 65
   5.4. Water Treatment Options 67

Chapter 6: Maintenance and Troubleshooting 69
   6.1. Regular Maintenance Tasks 71
      6.1.1. Cleaning Gutters and Filters 73
      6.1.2. Inspecting Storage Tanks 75
   6.2. Troubleshooting Common Issues 76
      6.2.1. Leak Detection and Repair 77
      6.2.2. Addressing Algae Growth 78
   6.3. Winterizing Your System 80

Chapter 7: Legal and Regulatory Considerations 82
   7.1. Understanding Local Regulations 84
   7.2. Permits and Restrictions 85
   7.3. Best Practices for Compliance 86
   7.5. Safe Water Use Guidelines 87
   7.6. Environmental Impact Assessments 89

Chapter 8. Integrating with Existing Farm Systems 92

8.1. Assessing Compatibility 93

8.2. Irrigation Systems 95

8.3. Types of Irrigation Systems 96

8.4. Livestock Watering Solutions 98

8.5. Designing Watering Systems for Livestock 101

8.6. Optimizing Water Use 102

8.7. Efficient Irrigation Techniques 103

8.8. Drip Irrigation 105

8.9. Mulching and Soil Management 106

8.10. Benefits of Mulching 108

Chapter 9: Future of Rainwater Harvesting 110

9.1. Advancements in Technology 112

9.2 Climate Change and Water Security 114

9.3 Community and Policy Initiatives 116

INTRODUCTION

Rainwater harvesting is the collection and storage of runoff from rooftops, the land surface, and/or road surfaces. For domestic use, rainwater collected from the roof can be easily treated with a sediment filter and chlorine. Communities can also benefit from overcoming rainwater shortages during dry seasons, decreasing pollutant loads to local rivers, and reducing runoff velocities that erode riverbanks and adjacent land sites. The simplest rainwater collection system is a rooftop gutter that carries water to a barrel or cistern located at the corner of the house. More complex systems pump water underground and store it in lined ponds.

Rainwater is very low in salts and other dissolved minerals. Rainwater pH is usually close to neutral 7. This makes rainwater a very high-quality irrigation source for plants sensitive to salts and high-pH irrigation waters drawn from underground aquifers. Rainwater harvesting for agriculture generally requires large catchment areas and storage volume to satisfy the high water demand for crops. For example, a 2,000-square-foot catchment area (a roof about 45 ft by 45 ft) can provide about 1,000 gallons of water for garden irrigation after a 1-inch rainfall. Smaller systems can still provide significant water savings for small gardens. Communities in arid regions often supplement their municipal water supplies with powerfully treated pumped rainwater.

The Importance of Rainwater Harvesting

Rainwater harvesting at the household level is one of the oldest and most environmentally sound subsistence strategies, as well as one of the most promising remedies for insufficient and unsafe water supplies. The importance of rainwater harvesting has been highlighted even more by modern events like pollution of surface water and groundwater.

In many societies, little, if any investment is made in harnessing rainwater. People in the driest regions often distress when rain is unusually abundant, and the only response to those not dying of thirst is to migrate. By moving to another area, they make water consumption unsustainable in two places instead of one. Such communities will remain at risk of hunger and malnutrition as long as they depend on these erratic fluxes for water to grow food. If the resource base in these drought-prone areas is seriously degraded, they may also be at risk of hunger even if they capture and store rainwater.

Benefits for farmers

Rainwater is the purest form of water. Even when it passes through city pollution, it washes itself on the way to the ground and removes dirt from the air. By harvesting this valuable natural resource at your home, you can help protect the environment as well as help moderate the quantity and quality of rural surface water and groundwater. Those farmers who already have ponds or in-ground cisterns can fill them with harvested rainwater to increase their water supply for livestock or to use for lawn and garden irrigation. If farmers have a greenhouse, they can use harvested rainwater to water their plants, which is preferable to using hard water and is more cost-effective.

For homes that already have rain gutters, collection is as easy as placing a rain barrel under a downspout. For larger storage volumes, consult resources and companies that provide plans for the construction of underground storage tanks, or companies that sell above-ground storage tanks. Before you begin to harvest rainwater for use around your homestead, gauge and increase your awareness of the benefits and challenges of rainwater harvesting. Then you can decide what rainwater harvesting system will best meet your needs. No matter which system you choose, you will need to prepare your existing structures to receive harvested rainwater. This text focuses on the selection and installation of screens to keep debris out of the rainwater flow, first flush diverters to prevent pollution of stored water, and covers for storage tanks to prevent algae growth and mosquito breeding.

Environmental impact

Aside from the direct benefits of rainwater harvesting to specific individuals, communities, or commercial enterprises, the process of designing, constructing, and maintaining rainwater harvesting systems can have a general societal benefit in heightened awareness of water issues and possibly greater acceptance of other, less beneficial water conservation practices. Mentoring programs that involve helping people design, build, and use rainwater systems are popular and can produce both a higher success rate of implemented systems and higher awareness and acceptance of water conservation concepts.

Rainwater harvesting has a host of environmental benefits. It reduces the volume of stormwater in drainage systems (potentially reducing the need for new or expanded storm sewer systems) and reduces the likelihood of localized flooding. Because rainwater is collected, transported, and used near its source, energy is saved and air pollution is reduced. In some areas, concern for water quality in reservoirs has led to restrictions on the use of released stormwater to irrigate gardens and landscapes, which can be the largest single use of harvested rainwater. The same restrictions do not apply to rainwater collected from roof surfaces.

Economic advantage

Rainwater harvesting captures, diverts, and stores rainwater for later use. It provides water for landscape irrigation and, if properly treated, can supply water for nonpotable uses in the home. Harvesting saves money if a well goes dry or when using a public water system that charges by the gallon for water use. Harvesting also reduces demand on both sources, which can be important in some areas during dry seasons. Economic benefits increase when rainwater is used for multiple purposes and when households help design and install systems, reducing costs for professionals.

For many households, the economic advantages of rainwater harvesting are the primary reasons to collect and store rainwater. These families often live in rural areas without easy or low-cost access to public water supplies. If their well goes dry, the cost to drill a new one can be prohibitive. Families in these areas pay close to $2-$5 per gallon to install new systems and even higher prices to purchase water from water delivery services. Urban homeowners have similar concerns about the cost of drilling a new well if their existing well goes dry.

Additionally, many urban homeowners now use public water supplies that charge by the gallon for residential water use. The more water their landscapes require, the higher their summer water bills. Rainwater harvesting can reduce these bills. Since both rural and urban areas may experience seasonal water shortages, reducing demand on available sources becomes an important consideration.

Chapter 1: Understanding Rainwater Harvesting

Why Harvest Rainwater? As people who are concerned with sustainable living, you are deeply aware of the need to conserve water, a precious and increasingly rare resource. By harvesting rainwater, you can reduce your dependence on municipal water supplies. Harsh governmental restrictions on water use during drought will become less of a threat, for you can store enough rainwater to get through the longest dry season. The stored rainwater can be used for many beneficial purposes, from watering your garden, landscape, or pasture, to washing your car, to providing potable water for your family and livestock. As an added benefit, rainwater is naturally "soft", so it can help keep your plumbing and appliances in good working order while saving you money on detergents and other chemicals.

In a nutshell, rainwater harvesting is a way to capture, divert, and properly store the rain that falls on your land, for beneficial use. Historically, rainwater harvesting was how humans provided themselves with life's necessities of clean, fresh water for drinking, agriculture, and survival. As urban centers developed, the natural supply of water began to be polluted and depleted, so that city dwellers had to capture and store rainwater merely to survive periods of drought. As technology advanced, water supplies were developed from rivers, lakes, and aquifers. Yet, even today, most of the United States faces periodic droughts, when municipal water supplies are taxed, and farming operations can come to a standstill. On a more modest scale, nearly every homesteader or gardener has experienced the hardship of wilting plants, and withering crops, due to lack of a reliable water source.

1.1. What is Rainwater Harvesting?

Rainwater harvesting is considered to be one of the best ways to utilize rain in regions where rainfall is seasonal. It typically involves the collection of rain from a roof, which is then filtered and stored in a rain barrel, cistern, or pond for later use. This stored water can then be used for irrigation of gardens and lawns, for animals to drink, and in some cases for household use. The key to maintaining a good supply of harvested rainwater is to make efficient use of every drop that falls - to conserve water in the storage, distribution, and use processes as much as possible. With careful planning and the use of appropriate materials and methods, rainwater harvesting can be an effective means of providing water for your homestead.

As we experience climate change on a global scale, many regions are faced with extended periods of drought and a marked decrease in available water for household and agricultural use. Rainwater harvesting is a simple and ancient means of collecting and storing water for future use. By capturing rainwater close to where it falls and storing it in natural or man-made reservoirs, we can provide a sustainable source of water for our gardens and livestock. Rainwater harvesting reduces reliance on ground and surface water sources and can be especially beneficial in areas where aquifers are rapidly being depleted or where water restrictions are in place. Furthermore, rainwater is free of the salts, minerals, and chemicals that are present in other water sources, making it an ideal source of water for plant irrigation.

1.2. Definition and Overview

Humans have also been harvesting rainwater for thousands of years. One of the earliest known instances of human use of rainwater was the rooftop collection systems in ancient Rome, China, and the Middle East. People living in rural areas where municipal water systsems do not exist, or in urban areas during droughts or where poor water quality exists, may choose to harvest rainwater. Today, the need and desire to implement rainwater harvesting is growing. There are many benefits to harvesting rainwater at your homestead.

Rainwater harvesting involves the collection, conveyance, and storage of rainfall for later beneficial use. This is one process that has been happening for billions of years on Earth. The sun heats the Earth, water evaporates and condenses in the cooler upper atmosphere, eventually forming clouds, then falling as rain. The rainwater that falls on your roof, driveway, and other impervious surfaces has been harvested many times before it enters the storm drain or overflows into a creek. However, traditional rainwater harvesting is usually harvested at a big scale by water corporations or local councils, treated, and pumped through long distribution pipelines at high costs to the end users.

1.3. Historical Context

Rainwater harvesting has been practiced for more than 4,000 years, with the earliest examples found in the archaeological remains of the Mohenjo-Daro and Harappan civilizations, located in present-day Pakistan and northwestern India. These people used simple earthen pots and large storage jars placed beneath the eaves of homes to capture and store rainwater for drinking and cooking.

In ancient Rome, villas were built with roof terraces that drained into cisterns buried beneath the building to store water for domestic use. The existing roof and its supporting structure were designed to carry the load of the soil, thereby allowing the terrace to be used as a garden area.

From early civilizations to the present day, people who live in arid and semi-arid regions have relied on the collection and storage of rainwater to meet their basic water needs during the dry season. As human settlements expanded, localized collection systems evolved into community-scale cisterns, some of large capacity.

1.4. Benefits of Rainwater Harvesting for Farms

Lack of privacy diminishes human quality of life and the noise buffer effect of dense foliage or forests is the best barrier to intrusive sound. Providing and maintaining good privacy cover for homes and farm buildings often requires more water than is available from the natural drainage area. Even in periods of inadequate rainfall, water can be harvested from complementary hard surface runoff by proper water collection system design. For this type of harvest, it does not even have to rain at the site where water is needed, greatly increasing the value of the harvested rain. It is the absence of economic value in rainfall that allows neighbors to use each other's surplus water. Therefore, neighbors must agree on how to manage excess harvested water.

Despite rain's obvious importance in the functioning of nature and its necessity to all life, adequate rainwater can shield crops or buildings. Rainwater harvesting brings a delayed benefit to farming for no immediate visual effect occurs when Mr. and Mrs. Jones begin harvesting rain. Nevertheless, rainwater harvesting is beneficial to any farm, ranch, orchard, or vineyard and brings several advantages. Harvesting rainwater reduces erosion gullies, their debris, and damage down-slope which can choke reservoirs and clog irrigation ditches. It can lower or eliminate flash flood damage. Rainwater harvesting can increase groundwater recharge as well as reduce energy costs associated with pumping water from wells.

Chapter 2: Planning Your Rainwater Harvesting System

Once you have decided to install a rainwater harvesting system on your homestead, proper planning is necessary to ensure that you will meet your water goals. Planning involves analyzing your site and water needs and then designing a system that will meet those needs. Installing a system that is poorly planned is, at best, a waste of resources and, at worst, can lead to serious erosion, water damage to buildings, or contaminated water. This chapter will help you develop a plan that will lead to a properly designed system tailored to meet your water needs. After completing this process, you may find that you are unable or unwilling to execute a rainwater harvesting system; alternatively, you may discover that you need to install a larger tank or additional collection surfaces. It is better to discover that your system cannot meet your needs before installation, rather than during a drought when your cistern is dry.

Before Moving on These are What You Will Need:

- Measuring tape

- Graph paper

- Ruler or other means of drawing straight lines

- Pencil with eraser

- Calculator

The first step in planning any rainwater harvesting system is to determine how much water you will potentially use and for what you will use that water. This requires you to develop a water budget, a forecast of the amount of rainwater needed to meet your household's annual water demand. It is important to develop an accurate water budget to prevent the overestimation or underestimation of a system's ability to provide you with water year-round. While you can always supplement your harvested water with other water sources such as drilled wells or municipal public water systems, overestimation of the system's ability to provide water year-round could lead to unsatisfied water goals, increased soil erosion, or other negative externalities.

2.1. Assessing Your Water Needs

Annually, a single person uses approximately 18,000 gallons of water. For a family of four, this amounts to about 72,000 gallons. Thus, a basic consideration in planning for rainwater harvesting is to verify the amount of water that you will need and especially the amount of water that you can harvest on your site. It is also important to note that your actual water use can be influenced by a myriad of consumer appliances that use rainwater.

To roughly calculate your annual water needs, you would add up all the various needs of water in your household. The various needs might include drinking, cooking, bathing and washing, sewer discharge, laundry, dishwashing, fire protection, garden and lawn watering, and livestock and pet watering. In most homes, non-potable uses of water usually exceed potable use. Laundry, toilets, and dishwashing use large quantities of water but do not demand high-quality water. Providing rainwater for these uses will lower the demand for a potable water supply, if there is one, and make more efficient use of rainwater.

2.2. Estimating Water Usage

A key step in designing a rainwater harvesting system is to estimate the amount of water used and the demand for water throughout the year. Since water use varies greatly among individuals and families, there is no substitute for making a list of the water-using appliances in a home and noting how frequently they are used.

The total daily water use for each home can be estimated by summing the average water use of each appliance and adding some estimate of operation of other appliances. For example, the use of washing machines may be added to kitchen flows, the use of dishwashers may be related to kitchen use, bath use may be added to shower flow, etc. Take care to include all likely uses of water when making these estimates. Furthermore, water use is rarely evenly distributed throughout the day. Average water use during 24 hours may be quite low compared to maximum instantaneous use, which typically occurs in the morning or the evening, before and after work.

When estimating water use for a home, it may be necessary to compare the estimated volume of water used during these peak demand periods with observed peak flow rates from existing service connections in the area. If the observed flow rates at these times are significantly greater than the estimated water use for the rainwater harvesting system, the rainwater system may be undersized, unless one or more of the appliances can be disconnected from the municipal water supply.

2.3. Seasonal Variations

Rainfall amounts have large seasonal variations in many parts of the world. This makes the timing of rainwater harvesting different from the timing of water demand for many uses, such as irrigation and domestic use. The surplus of water is harvested when it is available and stored for use when needed. Domestic water use is fairly constant throughout the year. The storage requirement for harvested rainwater is therefore to a large extent determined by the demand pattern for domestic water use. If water demand is fairly constant during the dry season and the dry season is short, then water storage need not be large to provide sufficient water to bridge the gap between supply and demand. If the dry season is long, with a short period of very little or no rainfall, then the storage requirement for harvested rainwater may be too large to be economically feasible as a source for domestic water use.

Furthermore, rainwater quality is generally better at the beginning of the rainy season after a long dry period without rainfall than when rainwater is collected after a period of continuous rainfall. Since stored water is frequently used at the homestead for drinking and cooking, it is important to keep health risks to a minimum. Smaller storage capacity, using excess rainwater for other productive uses like vegetable gardening, in combination with a water treatment system for potable water at the point of use, maybe a more economically viable option, resulting in a more reliable and safe use of rainwater.

2.4. Site Assessment

To best use available water, make a thoughtful assessment of your site before designing and building a rainwater harvesting system. Key factors to consider include:

- ❖ Total annual rainfall and the rainfall pattern.
- ❖ Average and record high and low temperatures for summer and winter.

- ❖ Dominant overall weather pattern, e.g. long dry seasons or periods of high humidity.
- ❖ Local variations due to proximity of coast, city, hills, or exposure to prevailing winds.
- ❖ Amount of roof area available to collect rainwater, dimensions, and slope of roof.
- ❖ Proximity and elevation difference between roof, storage, and point of use, if any.

These and other factors you identify will influence system design and cost. Your assessment need only be as detailed as required for planning your rainwater harvesting system.

However, if you require a more complex system, you should consider more detailed aspects of the siting, design, construction, and maintenance of such features as:

- ❖ Infiltration structures, for example, trenches, dry wells, and French drains receive surplus roof run-off.

- ❖ Pumps and filters to move and treat harvested rainwater.
- ❖ Provisions to prevent freezing if applicable.
- ❖ Erosion control to protect the site and building.

Additional specialized information and advice may be needed to properly design and build such specialized features. The site assessment is primarily a planning tool. Use it to prioritize your requirements and make other decisions that contribute to a system that best uses available water at a reasonable cost.

Rainwater Harvesting System and its components

Chapter 3: Components of a Rainwater Harvesting System

Treatment components depending upon the desired end use of the harvested water, treatment components may be required to screen out debris and allow for plumbing and pumping of the stored water. Treatment components may include filtration, ultraviolet (UV) light, chlorination, activated charcoal, and reverse osmosis. Meeting standards for drinking water will likely include several treatment steps, while less stringent standards for water used for irrigation or toilet flushing will require fewer treatment steps.

Storage container: The storage container holds the harvested water until it is used. Storage containers, often called cisterns, can be made of a variety of materials, including concrete, metal, and plastic. The size of the storage container is driven by intended water use, the requirement to store water from one rain event to the next, and the space available for its installation. Cisterns should be sealed to prevent the water from evaporating and to prevent mosquitoes and other insects from entering and laying eggs.

First flush diverters: As rain falls on the catchment, it washes off debris and pollutants that have accumulated on the catchment since the last rain event. As this "first flush" water is relatively dirty, diverting it away from the collection container helps to improve the quality of water in the collection container. First-flush diverters are either automatic or manual and are designed to capture the first portion of runoff from the catchment. The first flush diverter then allows cleaner water to enter the cistern through the conveyance piping.

Conveyance system: The conveyance system consists of the gutters, downspouts, piping, and couplings needed to move rainwater from the catchment to the storage container. Gutters are attached to the lower edge of the catchment.

Downspouts are vertical pipes that carry the water from the gutters to the collection container. Conveyance piping can be either the downspouts or other vertical or horizontal pipes that direct the harvested water to the cistern.

Catchment: The catchment is the surface from which water is collected and concentrated to flow at an acceptable rate for collection into a storage container. Roofs are the most common catchments used in rainwater harvesting systems. However, catchments can be any surface from which water can be collected and conveyed to the cistern, including decks and unpaved areas.

3.1. Catchment Areas

The catchment area is the first component of a rainwater harvesting system. It's the area from which the water is collected and transported by gutters to the storage tank. The catchment area may be a roof, concrete, or asphalt surface, an unpaved area, or any other surface that can collect water. By far the most popular catchment area is a roof, and it is with roof catchment that this guide is mainly concerned with.

Catchment efficiency is influenced by the slope of the catchment area and its surface texture. Low-slope roofs are less efficient than steeply-pitched roofs. The best surface texture is provided by a new roof. Asphalt tar and gravel surfaces, as found on flat or low slope roofs, gradually roughen with age and become less efficient. The catchment area for rainwater harvesting should be as clean as possible. Efficient gutters are essential to transport the rainwater quickly to the storage tank before much is lost through evaporation. Small bits of gravel, dirt, and leaves can greatly interfere with the efficient operation of gutters, so the catchment surface should be kept as free of debris as possible. Since a roof catchment area may be several hundred square feet, it is usually necessary to use a screening device at the gutter outlet to prevent debris from entering the pipe that leads to the storage tank.

Roofs

The amount of water that can be harvested from a roof is directly related to the area of the roof and the amount of rainfall that is received. In Massachusetts, each square foot of roof will collect about 0.6 gallons of water for every inch of rainfall that is received. Therefore, a 1,000 ft2 roof will collect about 6,000 gallons of water for every inch of rain that falls on it. With an average annual precipitation of about 46 inches, this roof could collect about 276,000 gallons of water per year.

Most homesteads already have roofs on their buildings, so using roofs to collect rainwater is generally the easiest way to harvest rainwater. Roofs should be constructed from materials approved for drinking water or the water should be treated properly for drinking. Roofs constructed from galvanized metal, aluminum, glazed tiles, thatch, slate, and food-grade plastic provide excellent rainwater for toilet flushing, laundry, and animal watering without filtration. Unpainted or galvalume metal roofs and painted concrete or tile roofs can be used to provide water for the irrigation of food crops after appropriate prefiltration. Roofs with lead-based painted surfaces, treated with chemicals to control moss and algae, or exposed to heavily polluted urban air (such as within a few hundred meters of heavy traffic) should be avoided. Copper roofs can be used, but water must be treated to remove the copper if it will be used for irrigation, as copper can accumulate in the soil.

Ground Surfaces

If you live in an area with intense summer rains and have a roof surface that you are considering for collecting this rain, you will need to have a gutter with a downspout installed on that side of the roof from which you want the rain directed. Gutters can be purchased in lengths and cut to size, or you may install a homemade gutter. Commercial gutters are made of aluminum, galvanized steel, or plastic. Aluminum is less prone to corrosion than galvanized steel, but it is not as strong. Small-diameter plastic pipe can be easily installed as a downspout. Be sure to secure the pipe to the building so it will not be blown down in a storm. The plastic pipe can be connected to the ground storage tank or allowed to drain into a smaller tank located near the ground, or it can be directed to irrigate trees, shrubs, gardens, or other plants.

The forecast is normally calm and sunny before rain. Use the opportunity to allow your ground surfaces to dry out. Bare ground can seal over very rapidly, but if a crust has formed, this should be broken to permit water to more readily enter the soil. If you have a mechanical means of breaking the soil surface, do so. Otherwise, a rake or hoe will work. Ground surfaces that are dry when the rain comes will absorb and store the most water. If you can plan a planting schedule that will allow soil to dry out before expected rains, you can make use of this natural irrigation process for your benefit.

3.2. Gutters and Downspouts

Roofs without overhangs will require gutters and possibly some form of a gable extension to catch the water as it runs off the roof. Gutters come in several materials, including plastic, vinyl, aluminum, steel, and wood. They also come in different shapes or profiles such as round, square, or ogee (molded to resemble C-type crown molding). Most homesteaders choose plastic or vinyl gutters, because they are relatively easy to work with. These materials are available at building-material stores and can be cut with a hacksaw. Plastic and vinyl gutters are attached to the fascia board with spikes or screws inserted through a bracket that sandwiches the fascia board. A drop outlet is used to connect the gutter to the downspout. The most common size for homestead buildings is 2 x 3 inches, but 3 x 4-inch and 4-inch round downspouts are available, should the need for a larger downspout arise.

The drop outlet is inserted into a hole cut in the bottom of the gutter. Metal gutters must be soldered to connect the drop outlet, while plastic and vinyl gutters will use a connector that snaps onto the gutter and drops onto the outlet. Elbows are used to connect the downspout to the outlet, which can point up, down, or straight out from the building, depending on the location of the outlet and the desired location of the downspout. Metal gutters may require a downspout strap attached to the wall to support the weight of the downspout; plastic and vinyl gutters have brackets that attach to the downspout. The downspout should have an elbow at the bottom to direct the water into a drain at least 5 feet from the building. If the water is not directed away from the building, it will flow back into the foundation trench and the underground drain.

Materials and Installation

The primary elements needed for a rooftop rainwater harvesting system are simple, and people in the developing world with money and access to some building materials will be able to implement the technology. The most common materials are simple;

they include gutters or troughs, conveyance pipes or downspouts, tanks or cisterns, first-flush equipment, and surface skimmers. Gutters can be made from any number of materials including wood, metal, bamboo, thatch, and plastic. They are commonly attached to the roof eaves and shaped like a half-pipe to conduct the water to the downspout. As gutters can fill with debris, frequent cleaning is necessary. Downspouts are often made from PVC pipe or sheet metal and can also fill with debris. A mesh covering serving as a screen can keep leaves and larger particles from entering the downspout or conveyance pipe.

Conveyance pipes can be made from clay, socketed concrete, metal, or plastic. PVC pipe is commonly used to carry the water from the downspout to the cistern. Conveyance pipes can also be laid underground; however, this is more costly. Cisterns can be made from a variety of materials including concrete, blocks, bricks, ferrocement, metal, and plastic. The choice of material will depend on local conditions and availability. Plastic tanks are lightweight and easy to transport; they can also be buried underground. Ferrocement cisterns are more labor-intensive but are relatively inexpensive. Brick, block, and concrete cisterns cost more to construct but are long-lasting. The first flush system can be as simple as a pipe with a ball valve that releases the polluted water until it reaches a clean state. Multiple pipes can be used and connected to a valve box serving as the first flush system. The valve box can be constructed of concrete, brick, or plastic.

Maintenance Tips

There are a few things you can do to keep your rain barrel in good condition and ensure it functions properly. Most importantly, you need to look at the safety and sanitary features of your rain barrel to ensure that standing water does not become a breeding ground for mosquitoes or other insects. Make sure that the rain barrel is covered with a screen of some kind to prevent insects from entering the barrel. The screen material can be either metal or nylon and should have a mesh count of 10-20 meshes per inch to exclude mosquitoes. Screens should not be so fine as to become clogged with debris carried by the water.

You should also clean the gutters of your rainwater collection system regularly to prevent leaves and other debris from entering the barrel and inspect the screen on your rain barrel to ensure it has not become clogged with debris. Depending on the quantity and type of debris entering the rain barrel, you may need to periodically remove the screen and clean it. Cleaning the screen annually is usually sufficient for most installations, but check the screen monthly at first to determine the rate of debris accumulation. Use a mild bleach solution to clean the inside of the barrel and lid once a year to inhibit the growth of algae and kill any insects or insect larvae present.

3.3. Filtration Systems

The following characteristics describe self-assembled low-cost filtration systems. These are simple, compact, and light. The maintenance is quite easy. Given that some of these low-cost filtration systems sometimes need large areas, homesteads with such available areas would benefit most.

Since bio-sand filtration involves both biological cleansing and sand filtration processes, it may not always be fast enough for rapid clogging removal in heavy rainwater. Clogging of bio-sand filters, because of heavy rainwater, needs to be addressed. In this context, processes that include some form of filter media regeneration or partial media replacement would be advantageous. Slow sand filters are capable of handling heavy rainwater and have been employed for rainwater filtration. Regenerative slow sand filters and hybrid slow sand filters could use bio-sand principles and concepts with advantages.

Slow sand filters and bio-sand filters are some examples of the above type of filtration systems. Slow sand filters are compact, self-assembled, and low-profile filters. Slow sand filters can also be designed to be tall gravity-driven flow filters. Short gravity-driven slow sand filters are compact and easy to maintain.

Bio-sand filters are self-assembled, low-cost, point-of-use filters, in which both sand filtration and biological processes take place. Such filters are quite popular in developing countries for the treatment of water for drinking purposes. Keeping the above characteristics in mind, bio-sand filters are explored as an option for low-cost, self-assembled, low-profile, cost-effective, small to medium-scale rainwater filters.

3.4. Storage Solutions

The type of storage you have will determine what uses you can make of your harvested rainwater. The storage capacity required will depend on the intended use. If you are supplying water for a garden, then the water needs are consistent throughout the year and capacity for storage can be determined provided you know the average monthly catchment yield. If you are considering using rainwater for toilet flushing and laundry, then the capacity of storage has to be such that it can hold several weeks' worth of water because demand for water in toilets and for laundry is fairly constant, whereas the yield from rainfall is highly variable and 2-3 months of dry weather can occur.

Various storage options are available. Small above ground storage tanks of less than 5000 litre capacity can be sited close to the building. They can be made of plastic, fibreglass, galvanized steel, or aluminium. Plastic tanks are commonly available and to be satisfactory must be of food-grade material. They are relatively cheap and are light in weight. Fibreglass tanks are also light in weight and do not deteriorate with age. Steel and aluminium tanks can be expensive, are more difficult to install and maintain, but are aesthetically pleasing. Underground storage tanks can be simple concrete tanks that are buried in the ground to save space. They can be made watertight by rendering the inside surface with a waterproof mortar. Commercial concrete septic tanks can also be used and are available in sizes up to 4500 litres. They are usually provided with an access cover at the top and can be extended to increase capacity.

3.5. Distribution Systems

How you get the water from the storage to where you will use it is the next consideration in the system. How to distribute stored rainwater around your property is determined largely by how you intend to use it. Fields and gardens may be watered by the gravity flow from outlet pipes from the tank or cistern. Water may be led directly to the plants you wish to water by the use of a perforated drainage pipe, or field ditches may be used.

Check valves, or a simple swing-type valve, placed near the outlet from your storage tank or cistern will help keep water from flowing out should the distribution line break, or if you have to remove a pipe for repair. A gate valve is necessary if you have to empty the cistern for any reason. Small plastic gate valves are inexpensive and are used on some small cisterns. Most drip irrigation systems require a pressure of at least 8 pounds per square inch in the distribution line. Not all require this much, however, so a system suited to your cistern or tank can be designed accordingly. Whether your distribution system will require a small pressure pump, such as a demand-type diaphragm pump, or whether gravity flow from the tank outlet will be adequate, is a matter of choice when designing the system.

How to design and build Your System

Chapter 4: Designing and Building Your System

If you have decided on the size of the cistern you would like to build, you are ready to find out how to design and construct it.

4.1. System Design Principles

Rainwater harvesting is an age-old process, steeped in respect for water and soil, and practiced in countless diverse cultures. It is a simple, effective method providing significant benefits in terms of water supply, erosion control, and water quality. To maximize these benefits, a rainwater harvesting system should be designed to suit the unique and specific conditions of the homestead, its residents, and its water demand. During the initial planning stage of a rainwater harvesting system, the designer must consider several factors and use design principles that are conducive to the overall function and performance of the system.

Regardless of the size or complexity of a rainwater harvesting system, the following design principles should be adhered to ensure system performance and reliability:

1. Use appropriate quality roofing materials.

2. Design the roof for rainwater collection.

3. Install a wire or mesh screen at the gutter inlet to prevent debris from entering the system piping.

4. Use gutter and piping materials that are easy to clean and will not corrode or deteriorate.

5. Use a first flush or diversion device to prevent debris and contaminants from entering the storage tank.

6. Size the storage tank appropriately.

7. Use a secure cover on the storage tank to prevent mosquito breeding, evaporation, and algae growth.

8. Install piping with proper slope and alignment.

9. Protect piping from freezing. 10. Provide tight, leak-free connections between piping components.

By following these design principles, the homeowner will ensure that the rainwater harvesting system functions as intended and provides clean, safe water for non-potable uses.

Efficiency and Sustainability

To be efficient and sustainable, a rainwater harvesting system should be designed and installed as a "closed system," i.e. the rainwater should be used and re-used for a specific purpose and not wasted. A closed system is more likely to be sustainable if the demand for water is less than the available harvested rainwater. Any demand exceeding the supply of harvested rainwater may lead to the use of pumped groundwater. If used for drinking water, this can lead to system contamination and increased risk to human health. Treatment of harvested rainwater to potable standards is expensive and not normally feasible for domestic supply except in emergencies. Additionally, exceeding demand and/or constructing a very large capacity system with resultant cost may reduce the savings associated with a system designed to reduce peak flows or to be a "zero discharge" system.

As some homesteads use large volumes of water only seasonally for activities like canning or cooling, it may be more efficient to design systems that collect and store enough rainwater for everything except seasonal high volume activities. The use of rainwater for irrigation of gardens and fruit trees, except during the rainy season, can reduce water use activities will help to ensure a sustainable system if water use demand does not exceed the volume of water harvested. The system can then be sized so that irrigation water needs are met, and the stored water used for other purposes when rainfall is not possible irrigating plants that require regular watering, it is best to use stored rainwater for these activities as groundwater in the dry season, putting them at risk of drying up.

Safety Considerations

Operation of rooftop rainwater harvesting systems for homesteads is generally low risk. However, the safety of stored water must be managed by preventing children or pets from exposure to the stored water. The area around the storage tank must not become slippery when wet. In cold climates, the overflow of water can cause ice to form on the ground, creating a slipping hazard. It is very important to design, build, or modify the homestead's rainwater harvesting system – including any necessary protective structures or barriers – to ensure safety and security, and to minimize any liability associated with accidents.

In addition to protecting the integrity of the stored water, maintaining the safety of the rooftop collection area is essential. This includes ensuring that the plumbing from the collection point to the storage tank does not pose a trip hazard, and that water does not leak and create a moist environment that could support algae growth. Preventing children or pets from gaining access to the stored water is also important for safety as well as health reasons. Some liability risks may be associated with accidents involving roof access, plumbing, or water storage, so it is recommended that safety be addressed with some urgency – especially after completing any new installation or modification.

4.2. Step-by-Step Installation Guide

Level a dry and sunny site of at least 200 square feet for the tank. The site should be on the downhill side of the home for gravity flow of water. Use a carpenter's level to make sure the tank site is level in all directions. Excavate a hole about one-half the depth of the tank and a diameter about one foot larger than the tank. This will allow access to the bottom of the tank if cleanout is ever needed and will allow for backfill around the tank to prevent floating.

Install a concrete slab if the tank will be placed on sloping ground to prevent shifting. Tank movement could damage downspouts and the house if tanks are not securely anchored. Build a roof to cover the tank if algae growth is a concern or if freezing weather is a problem. Algae growth can be eliminated by chlorinating the water in the tank. A roof over the tank may also maintain a higher water temperature and protect the tank from freezing.

Install a first flush diverter at the inlet of the tank to ensure that the greatest amount of roof runoff is collected. The diverter is a pipe slightly larger in diameter than the inlet pipe to the tank, installed horizontally with an elbow facing down. This design automatically allows the first few gallons of each rain to bypass the cistern. The amount diverted can be increased or decreased by changing the length of the horizontal pipe. Be sure not to collect runoff from chemically treated roofs, as the chemicals can contaminate the water and damage the cistern and water distribution system.

4.3. Common Challenges and Solutions

Rainwater harvesting systems for homesteads, especially those used for domestic purposes, often encounter several challenges. Some of these are large fluctuations in demand between wet and dry seasons, very little or no runoff during long dry seasons, high initial investment costs of properly sized storage systems, and limited space for storage. Overcoming these challenges often necessitates the adoption of innovative design and management solutions. Some of these, such as the incorporation of alternate sources of water supply, are briefly described. Other challenges that have been encountered in the implementation of rooftop rainwater harvesting systems in a few Asian countries are also briefly described together with the solutions that have been adopted in those situations.

Meeting large seasonal and dry weather demands with a rainwater harvesting system designed primarily for water savings results in very large storage volumes. Because the cost of storage is a significant component of the cost of the system, it may turn out to be unaffordable. Such demands may be met by increasing the size of the catchment and storage, but this would also increase the cost of the system. Investors may not be willing to spend beyond a certain amount. Satisfaction of such demands in a cost-effective manner may therefore require an innovative design or the adoption of another water supply source, such as pumped groundwater or surface water from a spring or pond, river water, or water from a municipal system.

Water Quality and Treatment

Chapter 5: Water Quality and Treatment

Just as with any drinking water source, the water you harvest from your rainwater collection system should be tested for quality on a routine basis and any necessary water treatment performed. Potential contaminants can be either environmental or from the catchment system itself. Environmental contaminants include microorganisms, animal waste, insects, and plants, along with chemicals from air pollution. Chemicals from roofing materials and other components of the catchment system can also contaminate the water. If tested, detected, or if you just want to be proactive about treating your harvested rainwater, several treatment steps can be done, including filtration, settling, and disinfection with chlorine or ultraviolet (UV) light.

Treating rainwater may not be necessary if the catchment system and its components are constructed of materials that do not contribute contamination to of the water and you live in an environment that has clean air quality, far from industrial areas. In some cases, rainwater is cleaner and softer than other water sources, but it should still be tested periodically. The catchment system should not be located near fertilizer use, heavy traffic, industrial operations, or wood-burning stoves. It also should be located away from bird, insect, or vermin infestations that could introduce microbial contamination into the system. The catchment area should be kept clean and free of leaves and other debris.

Rainwater may cause premature failure of plumbing components that it comes in contact with, such as copper pipes, because it has a low buffering capacity and can be slightly acidic. If rainwater is being utilized for irrigation, buffering the water with a suitable material may also be needed so that the acid rain does not cause damage to the soil, especially if the soil has a low pH.

5.1. Ensuring Water Quality

Rain is generally clean. As it falls, it may pick up a little dust, pollution, or other particles, but not a high concentration. However, after rainwater lands on your roof and flows into your gutters and through your downspouts into your cistern, it has the potential to become very contaminated for several reasons.

- Roofs will invariably have bird droppings on them. Bird droppings are known to carry several diseases. If a large flock of birds is roosting in trees over your house, consider this source of water unacceptable. Another source of fecal contamination is from small animals such as squirrels or raccoons if they have access to your roof area.

- Roofs can also have molds growing on them. If the roof is disintegrating, small particles of mold can be carried into your cistern. This is only a problem if you are allergic to mold.

- The main concern for roof water quality is chemicals. If your roof is made of asphalt shingles, it sheds small particles of asphalt into the water. The asphalt contains several chemicals such as polycyclic aromatic hydrocarbons, phenol, and chromium. These chemicals are harmful if ingested over a long period. This is the reason some experts recommend not using asphalt shingled roofs for potable water or not using potable water for vegetable gardens if the roof is asphalt. If your roof is new, you might want to let several rains pass before using the collected water.

5.2. Testing and Monitoring

Upon establishing your rainwater harvesting system, one of the first orders of operation, after first flush diversion and filtration, should be testing the water. Ideally, you should have an understanding of the annual cycles of your system and test the water at various times during the year. At a minimum, you should have some idea of the quality of water that you are using to irrigate your plants, bushes, and trees, and the surface area of the harvested water, after treatment, during late spring and summer.

Several types of home water testing kits are commercially available. Selection of the appropriate test kit depends on what you want to test for. Some test kits can test for several items such as contaminants and beneficial management materials. For a basic check of what you are irrigating, you should test for pH, electrical conductivity (measures dissolved salts in the water), and dominant ions (since rainwater and any potential harvested water will have a low ion content, the appearance of dominant ions can be an indicator of water quality). If you have other sources of water on the homestead, you may want to test for fecal coliform and other possible contaminants of concern. Be sure to read the instructions of the test kits you are using as they can vary greatly in methodology.

5.3. Common Contaminants and Solutions

One may reasonably question the quality of rainwater; after all, rainwater forms when water vapor in the air condenses around dust particles, becoming heavy enough to fall to Earth. Indeed, rainwater is typically of higher purity than other water sources; as long as the rain is not acidic, it should be free of a large class of chemical contaminants. Many environmental scientists collect rainwater for use in analyzing pollutant levels in the environment. Unfortunately, rainwater quickly picks up whatever is on the surfaces with which it comes in contact. Common contaminants from residential roofs and associated plumbing (gutters, downspouts) include:

- Heavy metals – lead (Pb), from lead flashing and paint, which was used in the past and may still be present in older homes. Zinc (Zn), which is a common roof cladding material, usually finishes high on the list of potential contaminants.

- Organic compounds, such as pesticides and PAHs - Microorganisms, which can be controlled through methods described previously

The first-flush diverter is probably the simplest to understand – and also one of the most effective. When the rain starts, the initial flow is diverted into a separate container by a valve or other mechanism. The first water that flows through the downspout washes off whatever is on the roof; it is sent to waste. After a short delay, the valve opens, and the "clean" water flows into the cistern. Depending on the system layout, this "flush" water may be diverted back to the downspouts after the cistern is full. This water has some advantages, particularly for garden irrigation, so it's not always wasted.

5.4. Water Treatment Options

The previous sections of this book have discussed water quality issues associated with using rainwater, found some common contaminants, and recommended some generally suitable methods for prefiltration, roofing and guttering. This section discusses treatment options in more detail. Treatments used for rainwater usually are regular maintenance of the system, first-flush devices, screens, sedimentation, and filtration, and disinfection as needed. Chemical disinfection is best for killing bacteria, viruses, and some protozoa. Boiling and pasteurization are effective but more energy-intensive. Ultraviolet light (UV) and ozonation are effective but generally more expensive methods of disinfection. Solar disinfection is best used for emergencies. No singular treatment will remove or inactivate all possible contaminants. However, the use of a combination of treatments described in this section will result in making rainwater safe for most uses.

Options for Rainwater Harvest and Use: The quality of harvested rainwater generally is very good. However, some sources of contamination are possible. These sources are associated with the roofing and gutters selected for the collection, the catchment area design, and the method of conveyance of water from the catchment area to the collection tank. All water should be disinfected before drinking or cooking with it because rooftop contaminants can represent a health hazard. Water should be treated to remove all particles before it is treated to disinfect it. Preferred methods of disinfection are chemical, ultraviolet light, boiling, or pasteurization. If a series of treatment steps is used, the water may be safe for drinking and cooking as well as washing fruits and vegetables and making ice. It is recommended that water used for these purposes come from a separate distribution system apart from the one used for human and animal drinking water.

Maintenance and Troubleshooting

Chapter 6: Maintenance and Troubleshooting

Rainwater harvesting for domestic use is very easy to maintain. Regular inspection and cleaning of gutters, downspouts, and storage tanks will help prevent problems. There are few maintenance procedures but many possible problems. This chapter lists many possible problems along with the troubles reported. It also lists the cause and solution of each problem if it is known. If you have a problem that is not listed, and you solve it, please report it. Your contribution will help us all.

Screen problems: Leaves, twigs, or other debris on the intake screen cause reduced flow. **Solution:** clean screen.

Mosquitoes or other insects in stored water cause disease risk. **Solution:** use mosquito netting or cover, or use insecticide.

Algae slime on the screen causes reduced flow and bad taste. **Solution:** clean the screen and remove algae from the water by filtering through sand.

Tar from the roof on the screen causes reduced flow. **Solution**: do not locate the screen above the downspout from a flat roof.

Gutter problems: Branches, leaves, bird nests, or other objects in the gutter cause reduced flow. **Solution:** clean the gutter, and block birds from nesting.

Gutter sags cause overflow at low spots, and roof or ceiling leaks. **Solution**: install gutter supports.

Gutters too short, or no gutters cause roof water to fall outside the catchment area. **Solution**: install more gutter, install gutters.

6.1. Regular Maintenance Tasks

Regular maintenance of rainwater harvesting systems is necessary, and many tasks should be performed on a scheduled basis. Several maintenance tasks should be performed to keep the various system components working correctly and to prevent clogging of the conveyance system or contamination within storage and treatment units.

In wet systems, the first flush device will trap mosquitoes and other insect pests drawn to the water, which can then be washed away at a convenient time. Wet systems may require more frequent inspection and maintenance because standing water in the collection and conveyance system structures can provide a breeding habitat for mosquitoes and other insects. Some maintenance activities will be weather-dependent, but specific tasks should nonetheless be performed during the dry season, the wet season, or whenever they can be scheduled.

An important maintenance task for the building roof is the removal of accumulated debris. Questionable material should be removed from the roof surface before the next rainfall. Easily removable leaf debris may be swept by hand. Care should be taken to reach over the edge of the roof while working and leaning a ladder against gutter supports rather than against gutter material. For taller buildings, roof cleaning may require professional services or the use of specially designed equipment. A smooth, unobstructed flow of water from collection surfaces to the storage unit is essential, and maintenance of the collection conveyance system is performed to prevent clogging, leaks, or overflow. The collection conveyance system includes gutters, downspouts, and, where required, above-ground conveyance channels.

6.1.1. Cleaning Gutters and Filters

Regular checks for debris are also required at the gutter, downspout, and filter. Typical debris found at the gutter and downspout consists of leaves, dirt, needle-like dry organic material from certain trees, and stones. Cleaning may be performed by hand. If the homestead is equipped with a dry system, cleaning can be facilitated by using a trowel or a small shovel. Filter cleaning needs to be done carefully and gently, to not puncture or damage the filter material. Use a soft bristle brush to remove debris from the filter surface.

If your system uses a first-flush diverter, this will also need to be maintained. The first flush diverter keeps the first concentrated flow of roof runoff from entering the storage tank. It is typically equipped with a ball and seat float mechanism, which after the water flow stops, seals off the diverter. Cleaning the ball, seat, and float, and the proper function of the first flush diverter should be checked only every 6 months or if there is visible sediment on the ball, seat, or float. Be sure to read the rainwater harvesting system operation and maintenance manual for proper procedures and safety precautions before performing any maintenance.

6.1.2. Inspecting Storage Tanks

Water tanks should be inspected at least twice a year to detect and correct problems early. Ideally, tanks should be inspected seasonally since problems can arise and be more easily detected and repaired when tanks are not full of water. Regular inspection also protects the water quality and quantity stored in the tanks. A pre-use inspection is often performed right before the beginning of a new system use. This type of inspection ensures that the system, especially if it has had a period of inactivity, is still in functioning order and that no new damage or wear and tear has occurred. More frequent routine inspections are performed while the system is in use. These inspections are intended to discover and correct any problems early as well as to validate and maintain water quality and quantity stored in the tanks.

It is a good idea to perform a structured tank inspection when the tank is being used in a rainwater harvesting system. Begin by looking around the tank for any obvious signs of trouble, such as wall bulging, leaking fittings, or damaged or pooled water near the tank's foundation. Next, determine the water level in the tank. If it is not visible, gently tap the tank near the top to determine if it is full.

6.2. Troubleshooting Common Issues

Here are short descriptions of a few common but easily remedied problems that owners might encounter with a rainwater harvesting system. Short cycling (tank level falling and pump activation, followed almost immediately by pump deactivation) can result from the air in the pressure tank, a defective pressure gauge, or an incorrectly wired float switch. An up-flow filter that is not periodically cleaned will reduce flow. Algae growth in the tank can be eliminated with a small amount of bleach or copper sulfate. Leaf and insect buildup in the first flush diverter can be alleviated by cleaning or by installing a screen over the inlet. A wet roof can be very slippery; it might be necessary to restrict roof access when it is raining to prevent a fall.

These are preventable malfunctions that may occur with the components of a rainwater harvesting system. Short cycling of a submersible pump can be caused by:

- Air in the pressure tank.

- A defective pressure gauge

- A float switch that is incorrectly wired

- A water hammer

- A prolonged very low pump flow

- And clogged water filters.

6.2.1. Leak Detection and Repair

If a leak is suspected at any point in the system, several detective techniques can be employed. It may be necessary to gain access to concealed portions of the pipe, such as by opening a wall or cutting a hole in the ceiling below the suspected leak. Signs of water leakage include staining and blistering or bubbling of painted or wallpapered surfaces. Water that wicks into nearby walls may also cause baseboards to swell. If the leak is occurring from a pressurized plumbing line, you may be able to hear the water escaping, especially at night when ambient noise levels are lower.

6.2.2. Addressing Algae Growth

Algae are present in the environment in various forms. They are microscopic plants that require sunlight to make their food through the process of photosynthesis. They are present in most water bodies and are considered to be the first link in the food chain that supports life in rivers, lakes, and ponds.

In rainwater harvesting systems, algae are also part of the natural succession in which plants colonize a habitat. Algae growth is encouraged and sustained by the presence of sunlight and nutrients (primarily nitrogen and phosphorus) in the water. If the rainwater in a storage tank appears green, then algae have been growing in it. This increases the demand for water disinfection since most algae are not killed by exposure to chlorine. However, algae are not harmful to water quality, and their growth indicates that the stored water is not chlorinated. Chlorine tablets should be added to the stored water to ensure a safe supply of water for domestic use if the water is not already treated.

To prevent algae growth, keep the storage tank dark. This is easily done since most tanks do not have windows. However, many tanks have roof hatches that allow sunlight to enter. If algae growth is a problem, cover the opening with an opaque material. Black plastic bags can be taped over the opening to exclude sunlight. If the problem is recurring, consider replacing the bag material with a hinged cover that can be opened for inspection and cleaning. A mesh screen should be installed under the cover to exclude insects.

Since the exclusion of sunlight prevents algae growth, the cover can be left off at night when the water is treated to disinfect it. This will allow the sunlight to enter the tank and facilitate the chlorine disinfection of the water.

6.3. Winterizing Your System

To winterize your rainwater harvesting system, first, drain all the water out of the tanks. Use this water in your landscape to provide some additional moisture to your plants before winter. If you have a first flush system, be sure to clean it out and dry it. Excess water in the first flush diverter can freeze and crack the plumbing. If you have an above-ground system or an above-ground pump, be sure to drain them. If there is any water left in the system, it can freeze and crack the plumbing. After you have drained the system, turn off the pump and disconnect it. Store the pump in a warm, dry place for the winter. Be sure to follow the manufacturer's instructions for pump storage. If your system has a filter, now is a good time to clean or replace it.

Be sure to winterize your gutters by cleaning them out completely. Leaves and debris left in the gutters can freeze and cause damage. If your region experiences snow and ice, install gutter heaters to prevent ice dams from forming. Be sure to document any maintenance you perform on your rainwater harvesting system and make note of any changes that need to be made for the next season. Properly shutting down your system for the winter will help prevent damage and ensure a longer life for your rainwater harvesting system. Be sure to check the weather regularly throughout the winter, and when spring arrives, check over your system to make sure it is ready to begin collecting rainwater again.

Chapter 7: Legal and Regulatory Considerations

One of the most frequently asked questions concerning the implementation of rainwater harvesting (RWH) systems at the individual or homestead level is: "Who owns the rainwater?" More accurately expressed, the question should be: "What laws, regulations, or doctrines apply to the ownership and use of rainwater?" The question of ownership or right to use is especially relevant where the capture, conveyance, storage, and use of rainwater may affect downstream landowners, waters of the state, or water that is being fully or over-appropriated in a basin. State responses to the question may vary, and currently, there are no state or federal laws that expressly address RWH at the homestead or individual level. It is clear, however, that in most states, the ability to capture, store, and use rainwater is subject to state regulation. State law, state constitutions, doctrines or regulations of specific application, and general water law principles may determine the use to which rainwater may be put.

There is no substitute for a careful review of applicable state laws, regulations, or rules, or, where state law is undeveloped, a consideration of state constitutions, water law doctrines, or general principles of water law. This chapter outlines some legal and regulatory considerations for a few states. The chapter does not set forth all laws, regulations, or rules, nor does it attempt to address each nuance or exception. These laws and regulations tend to evolve, and this evolution may lead to greater use of RWH or, conversely, greater restrictions on the practice. This chapter is not a substitute for individual legal research and consultation with a state-licensed attorney who has experience in this area.

7.1. Understanding Local Regulations

Rainwater harvesting from rooftops is an ancient practice that is regaining recognition as a sustainable source of drinking water when properly treated and stored. It may seem surprising that such a sustainable practice would be regulated, let alone banned, as has been the case in some localities. However, the mandates of legislators and village elders charged with the responsibilities of managing public health, safety, and welfare are often carried out through regulations put into place to ensure prescribed levels of water purity and distribution of safe water supplies.

Before investing in a rainwater harvesting system for drinking water or other domestic uses, it is important to begin by understanding national, state provincial, and local regulations. Sometimes the challenge is to determine whether such regulations exist at various levels, what must be done to comply with them, and who has jurisdiction.

7.2. Permits and Restrictions

Any changes to the plumbing systems require that the appropriate permits and inspections be undertaken. This may include a discussion with the plumbing officials to determine whether the rainwater system is going to be regulated as part of the municipal water system or as a separate entity. At the state level, the regulatory agencies want to ensure that rainwater systems do not affect the aquifer recharge system. Any overlap in the use of rainwater and the public water system should be appropriately regulated. The cross-connection rules are one example. Rainwater systems should be designed and installed according to the plumbing code requirements to prevent any contamination of the public water system. The public water system should be protected from any potential contamination from the captured rainwater that is being used for domestic purposes. Be sure that you are aware of the plumbing code, municipal code, and any covenants and restrictions for your area to avoid any surprises after the system is installed.

7.3. Best Practices for Compliance

Legal and regulatory compliance for rainwater harvesting in homesteads is currently best achieved by the following best practices:

a. **Prohibition of potable RWH** - To avoid potential conflicts with existing water rights, current formal rainwater harvesting regulations (or lack thereof) generally limit rainwater harvesting to non-potable use. The guidelines for rainwater harvesting for non-potable use in various local regulations, such as those of Texas, are considered very restrictive and caution the user not to implement potable rainwater systems without careful review of the applicable state regulations as it may not be allowed.

b. **Redundancy** - The Colorado Plumbing Code has included rainwater harvesting as an alternative method to obtain points to offset other prescriptive requirements but with certain conditions. The conditional nature of the permission has led to multiple precautions. Colorado has two areas where rainwater capture systems are commonly used. The first is in the mountains where potable water is secured from melting snow. RWH is used in the drier parts of the state as a source of water for gardens and landscaping. In both situations, a commercial system would be designed with redundancy and backups, such as a cistern overflow leading to a well that would be activated when the cistern is empty.

7.5. Safe Water Use Guidelines

Rainwater that is harvested and stored properly, meeting recommended procedures, is generally safe. The following can help assure safety:

a) Use only the freshest, cleanest water for drinking and cooking.
b) Water that is to be used for drinking should be boiled or treated with chlorine or another

disinfectant, such as hydrochloric drops available in drug stores (3-5 drops per liter of water and let it stand for half an hour before using it).

c) Protect stored rainwater from contamination by limiting access and handling, keeping containers tightly covered, and properly cleaning and maintaining the catchment area and storage container.

Normally, harvested water is used for purposes other than drinking or cooking without treatment. However, if water is to be used for these purposes, it should be treated to remove any disease-causing microorganisms or other contaminants. Water that has not been treated should not be used for drinking or cooking.

Several methods can be used to make rainwater safe for drinking and cooking. The best method is boiling water for one minute. Chemical disinfection can be used. Chlorine is highly effective but may not always be available, and it will impart a taste to the water. Additionally, chlorine gas is hazardous. There are other chemical disinfectants available, and they are usually easy to use. The water should be left standing for at least 30 minutes after adding the disinfectant. Boiling and chemical disinfection can be used in combination. The water can also be filtered before use. Any filter used should be capable of removing bacteria and viruses. Sand filters may be capable of removing some pathogens. Ultraviolet light can be used to disinfect water. However, it is generally effective if the water is free from suspended solids.

7.6. Environmental Impact Assessments

The objective of conducting an environmental impact assessment is the protection and maintenance of the natural, social, and economic environment that surrounds and could be affected by proposed projects or activities. An environmental impact assessment should identify, predict, and evaluate the natural and potential consequences, and adverse and positive effects of proposed projects or activities. With the identified series of legal and regulatory instruments, an environmental impact assessment could also ensure that a project proponent is sufficiently informed of all developmental prerequisites related to a proposed system of rainwater harvesting and can make adequate use of all proper protection and enhancement measures to ensure a sustainable economic, social, and environmental development outcome for the homesteads.

Environmental impact assessments specifically are legally envisaged for those activities that are not considered obligatory to have on hand a detailed project for construction, development, and the like before proceeding. Rainwater harvesting at homesteads encompasses a group of activities coupled with small-scale systems that are not subject to stringent specific building regulations, construction codes, or monitoring and enforcement control. As far as these systems rely upon the internal water circuitry of the homestead, the procedures of controlling surface water flows are applicable. In some jurisdictions, the building authorities undertake the responsibilities concerned with the construction and drainage of internal water systems. Nevertheless, the innovation and agency support of the whereabouts of rainwater catchment, storage, treatment, and reuse are not ordinarily subjected to detailed construction designs. This freedom may promote a wide range of risks, from unsystematic utilization of otherwise beneficial technology to inadvertent environmental damage, for example, by pollution of groundwater or deterioration of a nearby surface water ecosystem.

Chapter 8. Integrating with Existing Farm Systems

Existing farm systems can also be modified to take advantage of rainwater available for part of the year. Light structures that can be easily moved, like collapsible tents, can be used for drying harvested grains when rainwater is available. This reduces reliance on sunlight and thermal energy. Similarly, transparent plastic sheeting can be used to enclose a space for creating a greenhouse. In both these applications, rainwater can be used for some or all of the year for these farm activities. In dry regions, people sometimes forgo planting a vegetable garden if the available water is not sufficient for maintaining the garden through the dry season. Rainwater harvested during the wet season can be stored and used for irrigating the garden during the dry season, thus allowing a second cropping of vegetables.

8.1. Assessing Compatibility

Rainwater harvesting (RWH) systems for homesteads often have a low storage capacity to meet the year-round water demand for domestic use, thereby resulting in a need for seasonal scale efficiency. Although the primary concern of RWH is to facilitate household water self-sufficiency, additional embedded benefits can be gained in a compatible way. Assessment of the compatibility should thus invoke not only the various beneficial uses of harvested rainwater but also the water demand management (WDM) strategy of employing fit-for-purpose water for various demands through water-efficient appliances and behavioral change. The study revealed that significant societal benefits of RWH can be realized through a compatible approach.

Compatibility is important because it helps in the successful implementation of rainwater harvesting (RWH). The process involves checking the compatibility of the various components in the system (roof, conveyance, storage, treatment, and end-use) with each other and at the same time aiming to maximize all possible additional benefits that the system can provide, apart from domestic water supply. It was investigated if irrigation of a home garden and domestic micro-hydropower are compatible uses of harvested rainwater in addition to the household water supply, and also if water demand management through a fit-for-purpose approach can further enhance the benefits. The study was carried out for three homesteads differing in size and location. The sizes of the RWH storage tanks were optimized for the three homesteads to enable a fit-for-purpose approach during the rainy season. Then, simulations were carried out for the dry season to estimate the monthly water surpluses as a basis for identifying compatible uses of the harvested rainwater for the three homesteads. Furthermore, behavioral responses concerning water use and appliances suitable for a fit-for-purpose approach were considered.

8.2. Irrigation Systems

The water requirements for a homestead garden are small compared to farming or commercial operations. Watering a garden by hand from containers filled in a rainwater storage tank is a labor-intensive, but common method used by many small-scale gardeners. Connecting a drip or sprinkler system to a rainwater tank can greatly reduce the effort needed to water the garden. It also frees the gardener to carry out other tasks while the garden is being watered. Since rainwater is free of the salts and chemicals often found in wells or municipal water supplies, its use can increase soil fertility over time. This is especially important for gardeners who wish to utilize organic growing methods. With the rising cost of municipal water and concerns about water shortages, the use of rainwater in gardens is becoming an increasingly beneficial and practical option.

Rainwater harvesting can be integrated with drip and sprinkler irrigation systems that operate at low pressure. These are ideally suited to small-scale gardens typical of homesteads. Drip systems apply water slowly and precisely to plants. Water is released through small holes or emitters located close to the plants. Drip irrigation is the most efficient way to water plants, with little wastage of water. Sprinkler systems are less efficient than drip systems but are also effective in applying harvested rainwater to larger garden areas. Both systems can easily be operated from rainwater stored in a tank with a gravity-fed low-pressure outlet. A small pump can also be used to pressurize the system.

8.3. Types of Irrigation Systems

Rainwater can alleviate water shortages in the dry periods almost wherever such shortages are occasioned by an irregular distribution of seasonal water supplies. Several low-cost irrigation systems are compatible with rainwater harvesting. These include bucket irrigation, furrow irrigation, basin irrigation, and sprinkler irrigation.

Bucket irrigation is the simplest of all and requires the least amount of land. Small amounts of water are carried to the plants in buckets and poured onto the root area. This method is practical when the plants are close together and the supply of water is relatively small. Usually, plants that require a great deal of hand labor for both planting and harvesting are suited to this type of irrigation. Such crops include vegetables, tobacco, and some of the more valuable field crops during dry intervals at critical growing periods.

Furrow and basin irrigation require a somewhat larger land area. Water is collected from the storage area to the plants in small, shallow, hand-cut channels. This method is practical when plants are planted in rows and the land area is somewhat larger than that required for bucket irrigation. The rows should be relatively level. A slight slope is provided to permit wastewater to flow from the lower end of the row or basin.

Sprinkler irrigation is by far the most versatile of all the systems. It can be adapted to virtually any crop and any size of homestead. Water is piped from the storage area to various sections of the homestead. The piping is branched and positioned along the rows of plants, and at the proper locations, sprinkler heads are attached. When water is supplied to the system, it is sprayed over the plants. This method of irrigation is commonly used on commercial farms and is available in kit form for homesteads.

8.4. Livestock Watering Solutions

Rainwater harvesting at the homestead level is increasingly being promoted as an adaptation strategy to help households become more resilient to climate change, buffer the impacts of climate variability, and contribute to food security. In many semi-arid and subhumid areas, water demand is high, particularly during the dry season for livestock, domestic use, and small-scale irrigation. If rainwater harvesting systems do not have sufficient capacity to meet these high water demands, stored rainwater may become depleted before the next rainy season, thereby restricting the ability of households to use water for additional livelihood activities. This can lead to damage to water tanks as well as gully erosion near the tank location from unrestricted overflows.

Simple Solutions for Livestock Watering For decades, small-scale farmers and pastoralists have been using simple, low-cost solutions to meet the water demands of their livestock. In many cases, these solutions are more appropriate than attempting to use harvested rainwater for livestock watering. Common approaches include: allowing livestock to drink directly from harvested rainwater storage tanks; transporting water for livestock from an alternative source (e.g., a nearby river) and pouring it into the tank; using guttering to direct rainwater from the roof of a stable or barn into the water tank; constructing earth pan reservoirs near the homestead, which are filled by pumped water from the rainwater tank; and constructing small rock catchment dams in nearby seasonal watercourses.

In many regions, providing these small-earth reservoirs a short distance from the homestead increases the possibility of using rainwater harvesting to fill them, especially during the rainy season. Livestock can generally walk up to 1 km to access water, so this option is viable when the homestead tank water is used, and the rainwater tank or reservoir is dry during the dry season. Over the years, there have also been developed more sophisticated solutions, such as remote monitoring and solar-powered pumped water systems that can be installed at rock catchment dams, to ensure a raised water storage tank at the homestead is continually supplied with water for livestock, irrespective of the distance between the two locations.

8.5. Designing Watering Systems for Livestock

Rainwater harvesting from roofs is increasingly becoming popular, especially for livestock watering, as a sustainable water resources utilization approach. For small-scale homesteads, the livestock number is also small. This allows for small roof catchment areas to be drained into small storage facilities for serving as watering points close to the livestock pens. Water can easily be transported from large storage facilities to small tanks using gravity flow or portable pumps that can be operated using solar or human energy. This would help in minimizing the distance water needs to be carried for watering the livestock from the stored rainwater.

Ideally, each livestock pen should have a tank that is periodically filled from the larger centralized storage facility. The tank should have a small drinking outlet at a height suitable for the different types of animals, which are being watered from the tank. Water should be replenished in the tank from the larger central storage facility, without disturbing the animals when the water level in the tank becomes low. If this is not feasible, then the tank should have an automatic float valve to fill the tank as and when the water level in the tank decreases. For some of the animals, particularly poultry, water needs to be supplied at a location near the feeding troughs, which are placed inside the cages. Various piping (or tubing) arrangements with suitable connectors can be made to carry water from the storage tank to the various drinking points near the feeding areas of the poultry kept in cages. Small tanks can also be installed close to the cages, based on the size of the poultry unit, with each tank serving a specified number of animals.

8.6. Optimizing Water Use

Some consideration needs to be given to how best-harvested rainwater can be used in and around the homestead – not just for domestic use but also for productive use, i.e. for rearing animals, poultry keeping, fish farming, and small-scale gardening.

For domestic use, it is important to have water readily available - without the need to pump it up from a tank located some distance away from the house. However, for most of the time, this is not a requirement. Water can be pumped to an overhead tank (for gravity distribution within the house) as and when needed. For the tank to be located close to the house, consideration can be given to having it at ground level and building the house on a plinth. This arrangement would save on the cost of pumping the water to an overhead tank as well as on the cost of the tank's supporting structure.

8.7. Efficient Irrigation Techniques

The value of rainwater for crop production is highest when the water is delivered as a thin film directly to the soil surface around the plants. This is the only way to create favorable soil moisture conditions for plant roots without excessive evaporation losses. In areas such as the Mediterranean, where rain-fed agriculture is practiced and the summer dry season is long, a system that applies harvested rainwater as a thin film to the soil could allow for supplemental irrigation during the dry season, making double cropping possible in some areas. Consequently, it could greatly increase the productivity of rainwater on homesteads. Several efficient irrigation techniques could be used for this purpose. These include drip irrigation, subsurface drip irrigation, and the use of some types of perforated plastic pipes.

Drip irrigation is very efficient, but can be relatively expensive. It is well suited for crops grown in rows. Subsurface drip is somewhat less efficient than surface drip but has the advantage of not interfering with other field operations such as cultivation. It is especially suited to row crops that have deep root systems. Both surface and subsurface drip irrigation require low flow rates, which can usually be provided by gravity. The cost of a gravity-flow system depends on the topographic location of the homestead. However, even if a pump is required, the low flow rate for drip irrigation will allow the use of a small and relatively inexpensive pump.

8.8. Drip Irrigation

Linking rainwater harvesting to drip irrigation can greatly enhance the productive use of rainwater at the homestead level. It can be used in kitchen gardening, cultivating high-value crops, and tree plantations. There are specifically designed low-pressure systems for such use. Alternatively, in larger systems, a combined pump and filter unit can be installed at the water source to provide the required low-pressure supply. Drip irrigation has several advantages over traditional surface irrigation systems. These include higher water use efficiency, ease of operation, and reduced weed growth due to localized watering of plants.

A major constraint of drip irrigation is the generally high initial cost of the system. Developing appropriate low-cost systems for the homestead, including the use of locally available materials, could greatly encourage the wider use of this approach. Women members of the family, who are often responsible for kitchen gardening, could specifically benefit from such systems by gaining control over when and where to use harvested rainwater for irrigation. This could release stored water in larger tanks for other more critical drought-proofing uses.

8.9. Mulching and Soil Management

Mulching with hay, straw, and other crop residues can help reduce water loss and erosion. Mulches allow water to penetrate and be used by the crops while reducing evaporative losses from the soil surface. Maintenance of a good cover of living vegetation on the soil offers the best protection against soil water loss through evaporation. Other benefits of a living cover are that it recycles nutrients and helps control weeds. If the living cover is a leguminous crop, it can add nitrogen to the soil as well. As much as possible, grow crops in succession, alternating between rainy season and dry season crops, to keep the soil in continuous use. If the soil has not been used for some time, establish a cover crop to protect it from erosion and to recycle nutrients.

Soils under continuous cultivation with rainfed crops usually experience a loss of soil fertility. Soil organic matter levels can be maintained or increased by adding composted organic materials and other crop residues. Compost helps build soil structure, which in turn helps the soil hold more water. Compost also adds nutrients to the soil. Crop residues can be tilled into the soil, but this may accelerate the decomposition process to the point where nitrogen is temporarily immobilized, which could affect the following crop. If you are digging contours or planting basins, save the nutrient-rich soil from the upper levels and use it to refill the planting basin after heavy rains wash out the soil at the bottom. This will help prevent erosion and the loss of nutrients. In general, careful soil management in combination with rainwater harvesting systems will help ensure better crop growth.

8.10. Benefits of Mulching

Mulching has several benefits in rainfed agricultural systems. Firstly, it captures and holds in situ available water for subsequent crop use. Secondly, it provides additional water for plant growth by increasing the proportion of captured rainfall that can be transpired by established or volunteer vegetation. In the absence of mulch, a considerable proportion of rainfall may be intercepted by the standing dead material, especially if the standing residue does not quickly collapse onto the soil surface. Mulch captures this intercepted water and allows the standing residue to more readily fall over, thereby reducing the potential loss of subsequent rainfall. The collapse of standing dead material onto the soil surface also shields the soil from the impact of rainfall, reducing soil crusting and damage to germinating plants.

Thirdly, mulch can supplement soil fertility by allowing nutrients in the vegetation to return to the soil and decompose into the ground. This is especially important for nutrients that have translocated from the topsoil but are still within reach of plant roots. If not returned to the soil, the nutrient may be lost to the system. Fourthly, the increased biological activity supported by available water, nutrients, and organic matter can enhance pest control so that the rainwater becomes a significant additional hydrologic input for agricultural systems that can be actively managed to store flood flows. There may be opportunities to manage other geohydrologic benefits while the stored water is available for supplemental irrigation.

Chapter 9: Future of Rainwater Harvesting

It is estimated that by 2025, half of the world's population will be facing water scarcity. Water scarcity will lead to reduced food production and increased food prices. Integrated farm and household systems that conserve and make efficient use of water, such as rainwater harvesting, will become not only a matter of choice but of necessity. There will be an increased demand for expertise in rainwater harvesting. There are already many professionals, contractors, and organizations specializing in rainwater harvesting, but much more could be done. Research needs to focus on developing cost-effective and easy-to-build rainwater harvesting systems that are adaptable to various receiving surfaces and can be used for different purposes. Promising scientific and engineering developments, combined with modern communication technology, should make knowledge and materials related to rainwater harvesting widely available.

However, though rainwater is free and does not belong to anyone, there may come a time when rules and regulations may have to be implemented to ensure that those who need it most have access to it. At present, in many areas, high-water users are warned not to consume water during water restrictions while having rainwater in their tanks. The time may come when the installation of rainwater tanks is mandatory, as is the installation of cesspits and soak pits when connected to municipal sewer systems. This could be a way of giving back to the earth its natural capital, which we are depleting at an alarming rate, and guaranteeing that future generations will have at least the same standard of living that we have now.

9.1. Advancements in Technology

To promote and facilitate wider adoption of rainwater harvesting (RWH) for sustainable homesteads and communities, innovative advances in technology are occurring in a variety of application components. These include catchment technologies such as increased use of solar panels, a better understanding of contaminants and methods to eliminate them, use of smarter and greener materials like graphene, filters that are self-cleaning or regenerate easily with ultraviolet light, and ease of assembly; storage technologies such as better understanding of microbial activity in the water before use and methods to eliminate biofilms that build up in stored water, easier and safer underground tanks, and tanks that are stacked or collapsible when empty but heavy-duty when full; distribution technologies such as the use of small, simple-to-maintain pumps powered by falling water or low-cost solar power, and drip irrigation that is easily moved and adjustable; and use of Internet of Things technology for monitoring water use and availability.

Looking towards the future, more advanced technologies such as the integration of cyber-physical systems with Geographic Information System-based real-time weather information can be developed. This can further revolutionize rainwater harvesting to make it possible and attractive to implement smart and connected communities to have a reliable and sustainable local water supply. This chapter will focus on the current developments and how to implement the more advanced methods. The varying water demand for homesteads, based on the number of family members and animals, will guide the discussion throughout the chapter.

9.2 Climate Change and Water Security

In changing climate, weather patterns are in part a response to increasing atmospheric warmth. A warmer atmosphere adds to water cycle dynamics through more direct impacts of rising temperature. Climate change may increase the intensity and frequency of floods, droughts, tropical cyclones, and winter snowstorms. Anthropogenic climate change is the climate change that may be attributed to human activities that alter the composition of the global atmosphere and which, in addition to natural climate variability observed over comparable periods. These variations are changes in mean climate or the distribution of climate over periods, for example, ten years or longer. The climate is profoundly influenced by its variability on all time scales, just as the weather is profoundly influenced by its links to climate.

The twenty-first century has begun with a spate of extreme and minor weather and climate-related events, reminding humanity of its vulnerability to the vagaries of weather and climate. Climate variability, change, and weather extremes can impact both natural and human systems. One of the considerable challenges for the future will be to recognize that the two contradictory views, that climate is predictable on long time scales and that weather is amenable to short and medium-term forecast, are indeed consistent within the framework of near-surface climate as being a boundary value problem and that of weather as an initial value problem. It is important to realize the potential for using such knowledge to help address natural resource management, climate change adaptation, and weather risk mitigation issues, which are so critically important for sustainable human progress on this dynamic planet of ours.

9.3 Community and Policy Initiatives

In many urban areas, large community systems could involve capturing water from several institutional buildings such as schools, and storing the water in one or more centrally located tanks. The stored water could then be used for several purposes, including firefighting, street washing, and recharging public gardens or other common areas. Where a community system is used to capture water from several buildings, the plumbing is likely to be much simpler than for individual buildings, and cost savings on both the piping and the filtration system can be realized. This type of system may be more readily accepted by communities than several small individual systems. In some districts, local authorities have realized the benefits of such systems and have implemented policy initiatives to promote rainwater harvesting on institutional buildings.

Several NGOs, village committees, and self-help groups have promoted rainwater harvesting in communities and have developed their collective systems. The Roof Water Farming Program established by the Watershed Organization Trust in India has constructed several large roof water harvesting systems for local communities. For a cluster of villages, the program captured over 600,000 liters of water in 13 systems with a total roof area of 2270 square meters. The stored water was used for fish farming, and the excess was then released to recharge the groundwater. The neighboring village then pumped the recharged water for domestic use. The Water, Engineering and Development Center in Tanzania worked on a similar project with a group of 10 villages. Water was captured from 20,000 square meters of roofs and channeled into a series of storage tanks, from which it was used to fill fish ponds and for domestic use.

www.ingramcontent.com/pod-product-compliance
Lightning Source LLC
Chambersburg PA
CBHW050111230526
45470CB00004B/1774